Compost A - Z

All the lilliputian and gargantuan things about composting.

Mary Mauzy MPH, PhD & Jill Doyle

PHOTOS:

Page 38 reproduces the image released by the Environmental Protection Agency for the Wasted Food Scale that can be found at: https://www.epa.gov/sustainable-management-food/wasted-food-scale.

Page 8— Many thanks to WM (formerly Waste Management) for the landfill photo-ops and taking me onsite!

Thank you to all the wonderful contributors on Wikimedia. Without your contributions this book would not be so colorful.

Dickenson-Davis photography had taken the magnifying glass picture on page 31 and so graciously allowed its reproduction in this book.

Thinking Dirty
114 Cañon Ave
Manitou Springs, CO 80829
www.thinkingdirty.org

Table of Contents

Composting Challenge

We want to challenge each school in the world to do what BV Montessori did.

On average, 1/2 pound of food waste per child per day is coming from school cafeterias. Can you calculate how much waste could be coming from your school lunchroom?

Start at your school by talking to teachers, custodians, and the lunch staff.

Plan a waste audit to see how much compostable material and recyclable items are currently being thrown away.

Then start asking questions:

- Contact your recycling company to see what is accepted and if it has to be clean.
- Does your town have a company that will pick up compostable materials? Will you need to get approval to compost on site?

Did you know that kids throw away 87 pounds of cafeteria food in a year! If your school has 200 students that adds up to

8.6 tons of food waste per year!

Record your progress over a period of time and see how much trash you keep out of the landfill! Chart and graph the data. Can you beat Buena Vista Montessori?

Get more information at thinkingdirty.org; ask questions, get support, or kudos for a job well done!

thinkingdirty.org

Can you beat Buena Vista Montessori?

Before:
Four 50-gallon bags a day of cafeteria trash.

After:
One 25-gallon bag (or less!) a day of cafeteria trash.

Data is messy, just like real life! Record what happens and <u>always</u> tell the truth. Adjust charts according to the data.

In this case we had to combine all the non-landfilled compostable information together (milk, water, & compost) because the water started being poured into the compost towards the end of the year. Not a big deal, just be sure to explain and represent the data honestly!

BV Tiger's Cafeteria Waste over 83 school days

Date	Compost (lb)	Recycled (lb)	Milk (lb)	Water (lb)	Landfill (lb)	TOTAL Lunchroom Waste	Total waste (lb) diverted from landfill
1/14/19	17	1	6	3	47.5	74.5	27
2/21/19	25.3	2.6	11.3	13.6	31.1	83.9	52.8
5/17/19	35.4	8.3	7.3	*	3.4	54.4	51
						3271.4	
TOTALS	2069.4	353.1	712.5	489.5	2383.1	6007.6	3624.5

*composted

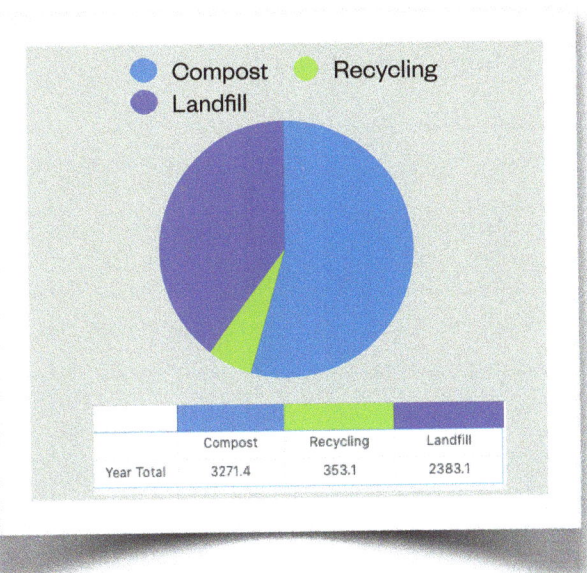

	Compost	Recycling	Landfill
Year Total	3271.4	353.1	2383.1

4

Everyone should compost.

There are many reasons why everyone should be composting, but here are a few of our favorites:

- organic material in landfills do NOT help decompose garbage.
- having the garbage bags/can/dumpster **not** smell is amazing!
- landfills create methane which is a factor in global warming.
- taking nutrients out of the soil and not returning any means higher inorganic fertilizer use.
- watching food scraps, lawn waste, and other organic material turn into awesome dirt is fun science.
- There is no waste in nature. The plants and animals do not have an 'away' to throw things or garbage trucks to pick it up.

Landfill myth: When I put compostable material in the garbage, I help the landfill.

Nope! The organic (compostable) material is trapped in layers of plastic in an anaerobic or anoxic environment at the landfill. Even if the material is compostable, the bacteria do not have oxygen for aerobic processes and will produce methane or it will be mummified in plastic. Landfills weren't designed to hold organic waste.

How to use this book.

We believe composting should be a part of daily life. Our goal was to create a book to start incorporate composting concepts into curriculums.

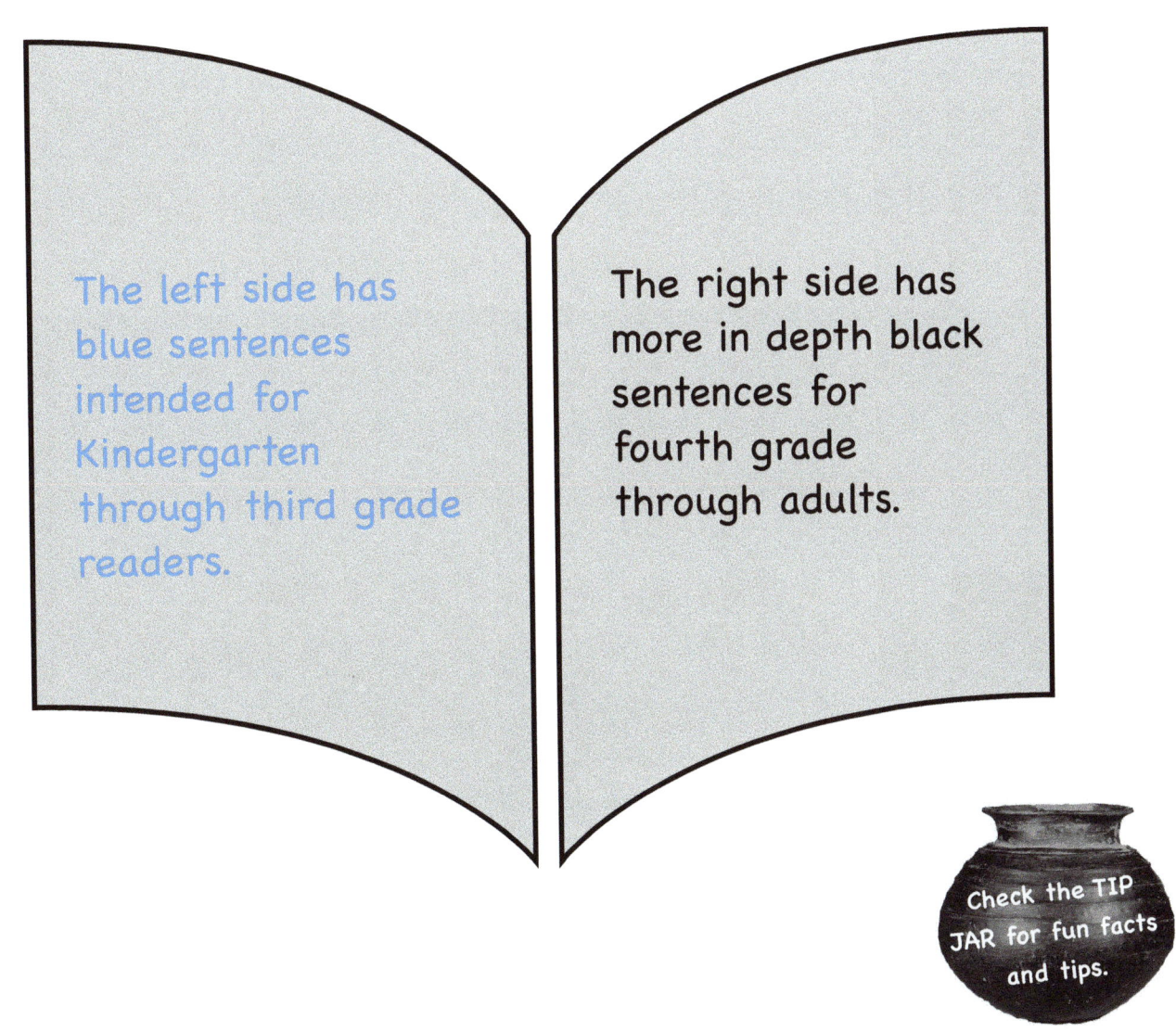

The left side has blue sentences intended for Kindergarten through third grade readers.

The right side has more in depth black sentences for fourth grade through adults.

Check the TIP JAR for fun facts and tips.

Aerobic & Anaerobic

We need aerobic reactions to produce carbon dioxide that plants can convert back to oxygen. Anaerobic reactions create methane which contributes to climate change.

Aerobic processes utilize the atom oxygen, whereas anaerobic processes do not. In science terms, respiration is the method used to make energy. Aerobic respiration in the compost pile uses oxygen (O_2) from the air to break down food waste and releases carbon dioxide (CO_2).

Unlike humans, microorganisms in garbage can 'breathe' in a different way when oxygen is not present. Microbes stay alive in plastic bags by switching to anaerobic or anoxic respiration. When microbes use respiration pathways without oxygen around, methane is 'exhaled' instead of CO_2.

Methane (CH_4) is a flammable greenhouse gas. Greenhouse gases trap heat from the sun and can contribute to climate change if too much is in the Earth's atmosphere. The atmosphere is a balance of different gases that protect the planet's surface from feeling the sun's rays directly.

Having a plastic bag over living things suffocates them. Using plastic bags suffocates the bacteria and fungus trying to break down our waste.

Methane is a molecule with one carbon atom attached to four hydrogen atoms. With ball-and-stick models, carbon atoms are grey or black spheres and hydrogen atoms are white.

Carbon dioxide is the gas released when you exhale.

Biodegradable

We need to use biodegradable products so we are not creating mounds of trash that will stay on Earth forever.

Even the Parthenon needed to be fixed because it degraded. All things will eventually degrade.

9

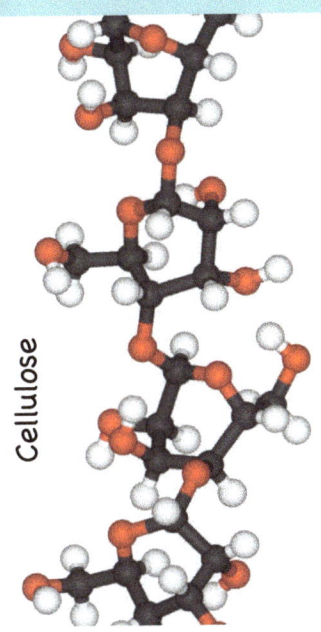

Cellulose

"Bio" comes from the Greek word for life. When something is bio-degradable, living things can chop it up and use the parts again. Food waste biodegraded by microbes forms pieces that are reused by plants and animals. Your digestive tract is a biodegradation zone full of microbes. The biomass, aka food, consumed gets broken down and turned into the cells that make up your growing body.

Here's the breakdown of some of the molecular terms:

Elements are the smallest particle that can be made using chemical reactions. An atom is a single element. Humans can't see a single atom of carbon, but a piece of charcoal is a visible element of carbon.

Cellulose is a macromolecule found in plant cells. Bacteria break cellulose into glucose.

When different elements bond together they are called molecules. When multiple molecules combine it is called a macromolecule. Carbohydrate is a fancy name for lots of sugar molecules chained together to make a macromolecule. Cellulose is the most abundant carbohydrate on the planet and is made of glucose molecules. Glucose is an important sugar that cells use to make energy.

Plastic and cellulose are both macromolecules made from carbon, hydrogen, and nitrogen, but plastics are not biodegradable. The chemical process used to make plastic bonds the carbon atoms together so strongly bacteria don't want to break it apart. Unlike cellulose being broken down into glucose molecules for cells to eat, when plastics become smaller pieces they are still just chunks of plastic. The smaller sized macromolecules of plastics are called micro-plastic or nano-plastic.

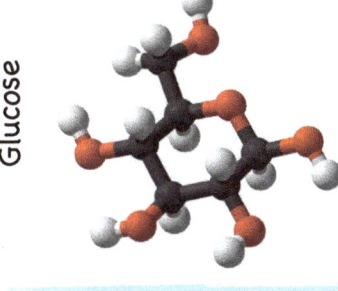

Glucose

These ball-and-stick chemistry drawings may look similar enough to us, but one is reusable to a cell and the other is not. Both are made with carbon, oxygen, and nitrogen (red), but the way the atoms are connected is not the same.

Glucose is the sugar molecule that makes up the macromolecule cellulose. Can you see the repeating glucoses?

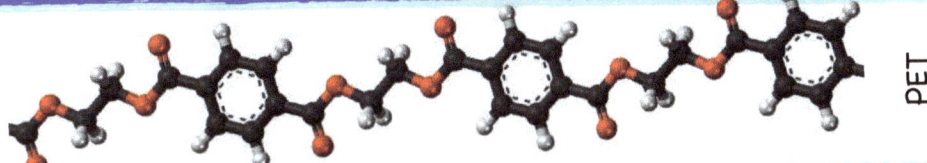

PET

This PET molecule is a plastic that organisms can not break into smaller, reusable pieces.

Carbon

Carbon in food waste can be reused by making compost. Items in the compost pile that look "brown" have a lot of carbon.

Dried yard waste and leaves are sources of carbon that people will happily contribute to a school composting program. Ask for donations during the fall and store the browns nearby until needed.

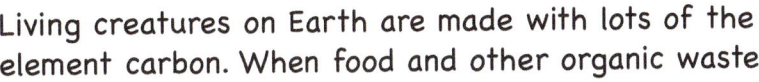

Living creatures on Earth are made with lots of the element carbon. When food and other organic waste aerobically decompose or compost, the carbon is released into the soil to be incorporated in new life. Material rich in carbon in the compost pile are called "browns." Examples of browns are dry leaves or other yard waste, sawdust, shredded paper, and cardboard. If your compost pile is smelly, you may need to add more carbon!

Shredded paper is easy to find and it will help keep the smell down if you have a lot of food waste or greens composting.

The amount of carbon that different browns have in them is not the same. For example, shredded paper has less carbon than sawdust. This means more shredded paper is needed than sawdust, by volume, to keep the compost from smelling. If size restrictions are an issue with the compost pile, go for carbon dense sawdust or wood shavings as a brown source.

As the waste decomposes the volume will decrease. You should notice as the process continues that your compost seems to 'shrink' as it finishes.

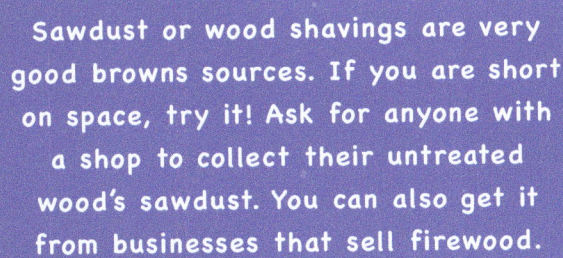

Sawdust or wood shavings are very good browns sources. If you are short on space, try it! Ask for anyone with a shop to collect their untreated wood's sawdust. You can also get it from businesses that sell firewood.

In the fall collect leaves. Store the leaves near the compost bin to have easy access to browns anytime.

Decompose

Mites, springtails, worms, and ants all decompose or break down banana peals, apple cores, and oranges. They can even break down the hard shell of an avocado.

Ever invite an ant to a picnic? I haven't either, but they are great at finding our spilled food. Ants and other insects see a compost pile as a feast!

Decomposition is the natural breakdown of organic matter. When biologists are talking about something being organic, they don't mean the label found at the grocery store. To a scientist or someone composting, the word organic indicates the material has carbon. In the forest or ocean, dead plants and animals decompose to supply nutrients, like carbon and nitrogen, back to the environment. This process can take months or years.

By learning from nature's methods to reuse nutrients, we have learned how to optimize decomposition for compost piles. Like decomposing, composting results in the breakdown of organic matter. The big difference is that composting can be completed in weeks instead of months or years.

The compost life cycle has three main stages.

Microbes first utilize the organic material.

Next, invertebrates eat the microbes and other organic matter.

Finally, the organic material becomes part of the soil again, usable by plants to start the growth cycle once more.

In the compost pile, combining browns and greens with water and oxygen makes the best ecosystem for quick decomposition.

These pine needles are decomposing beneath a Ponderosa pine. The most recently fallen pine needles are on top and still light brown. Underneath the older pine needles are black and breaking into smaller pieces.

Energy

The compost pile organisms need oxygen to function and break down the carbon to produce loads of energy!

Stir the compost to add oxygen. You don't stir compost with a spoon, but you can use a shovel to dig it out! Some places have so much compost they use augers or tractors to stir it.

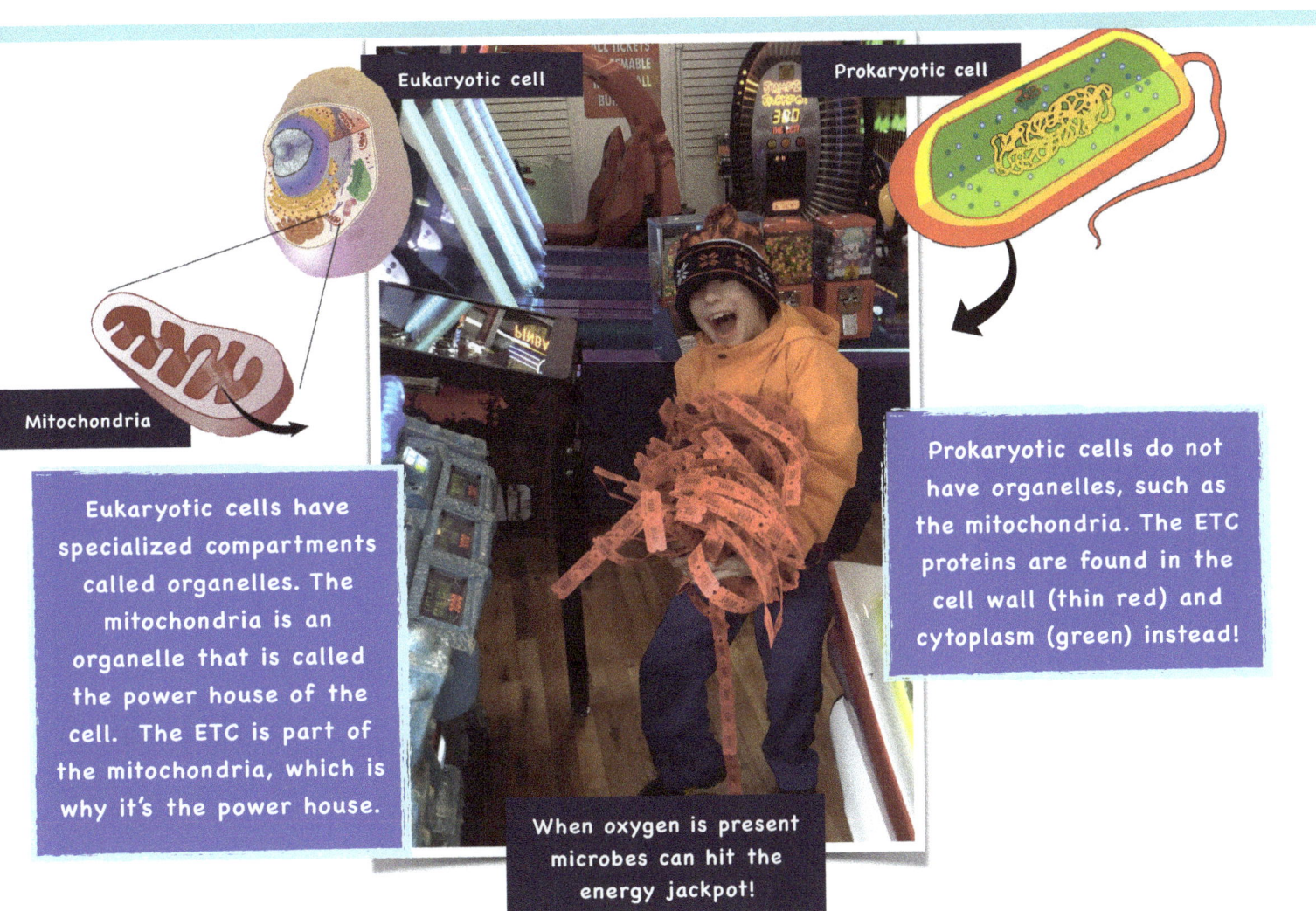

Eukaryotic cell

Prokaryotic cell

Mitochondria

Eukaryotic cells have specialized compartments called organelles. The mitochondria is an organelle that is called the power house of the cell. The ETC is part of the mitochondria, which is why it's the power house.

Prokaryotic cells do not have organelles, such as the mitochondria. The ETC proteins are found in the cell wall (thin red) and cytoplasm (green) instead!

When oxygen is present microbes can hit the energy jackpot!

All living things need to make energy to grow and replicate. If you eat something full of sugar, you tend to have an energy boost. Just like humans, microorganisms get energy from eating the sugar glucose. When oxygen and glucose are present, the microbe can have the equivalent of a sugar rush!

Cellulose (see B for Biodegradable) is made of many glucose molecules that microbes break apart like a candy necklace. The waste in compost is full of cellulose. Aerobic microorganisms use a series of events called the Electron Transport Chain or ETC to make a lot of energy with a single glucose molecule. The ETC is a series of reactions or steps. Similar to playing an arcade game and passing through all the levels, glucose, oxygen, and other molecules in the cell pass through the stages in the ETC to get to an energy payout. Without oxygen (O_2), the cell can't make it to the final level in the ETC. When a cell is able to use the ETC, it is like hitting the energy jackpot!

Fungus

Fungi are present to decompose materials back into the soil.

Mushrooms are a type of fungus that many people eat.

"Food is not garbage; it can always be food for something else."
~Barack Ben Amots

Flies and mushrooms are both eukaryotic organisms.

Fungus can come in many types: including yeast, mold, and mushrooms. Molds are a microscopic fungus that like to eat our forgotten leftovers. The mold will eat our food waste to make energy to live. When there are enough of the individual mold cells growing, these microscopic organisms look like a fuzzy mat. Although we look at moldy food as inedible for a human, putting it into the compost allows the mold, other types of fungus, bacteria, and invertebrates to have a nutritious meal. The compost that is created can be used to fertilize the next generation of plants. Herbivores can consume the plants, thus contributing to our food cycle.

Fungus are eukaryotic organisms that degrade food to make energy to replicate. Bacteria also degrade our forgotten food to make energy to reproduce, but a fungi has many different ways to replicate. Bacteria can only make a copy of itself, like a clone. Making a clone like bacteria do is called asexual replication. When a fungi makes a clone of itself, aka asexually replicates, it is called a spore. Spores are small and light so they can fly through the air and land on the next food source to grow. Fungi can also mix up their DNA to create a baby fungus cell, this is called sexual replication. Most eukaryotes cannot asexually replicate, which make fungus a pretty unique kingdom of organisms. Ask your favorite mycologist (person that studies fungus) more about these amazing eukaryotes.

This mold grew in a forgotten lunchbox. Can you see how fuzzy it looks? The black specks on top are the fungus' pouch of spores to release.

Mushrooms carry their spores under the cap, in the structures called gills.

Garbage

In a year, a typical family throws away trash that is the size of an elephant! We all need to decrease waste and throw away less to save the Earth.

An elephant weighs 7 tons.

Everything in a garbage bin is taken by a collection agency to a transfer station. The combined garbage at the transfer station is taken to a landfill. In Colorado Springs, one landfill receives 1,500 tons (that's 3 million pounds!) of garbage each DAY!

The average bag of trash from an American's house weighs 15 pounds. The average African Elephant weights 7 tons (14,000 pounds). When your family is throwing away an elephant a year, you throwing away about 18 bags of trash each week. Yikes!

This number may seem big, and it is, because it includes not just the garbage from home, but also what you throw away at work, at school, at the gas station, at the movie theatre, and everywhere!

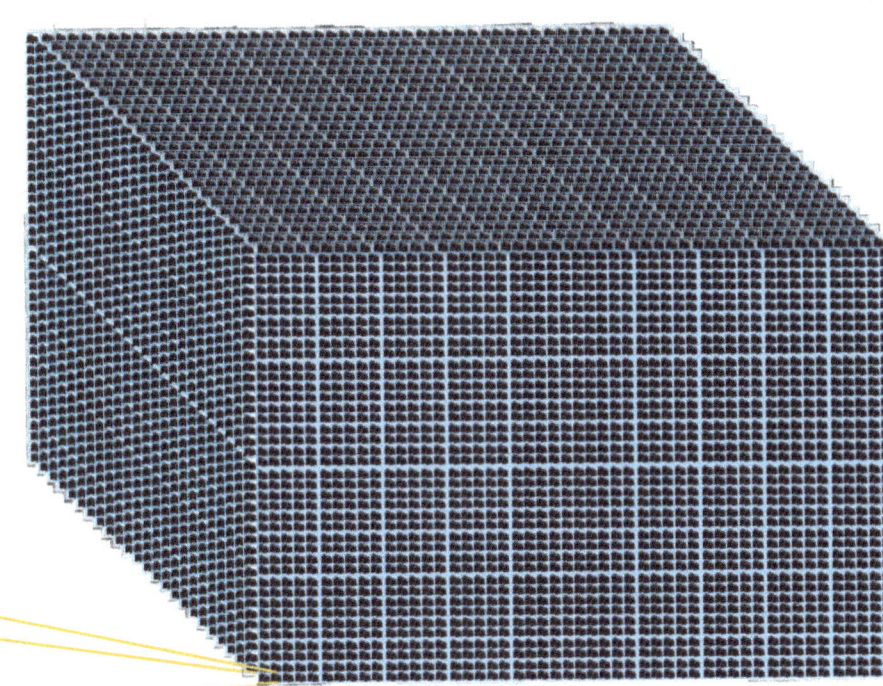

Family of 4's trash per week

Each bag of trash represents a 15 pound average bag of trash. This cube represents the 3,000,000 pounds of trash received daily in Colorado Springs Midway Landfill. The yellow box represents a family's daily trash.

Of the material ending up in the landfill, 75% is either reusable, compostable, or recyclable. Landfills are not interminable; once the space is filled with garbage more land is needed to hold our garbage. The companies collecting our trash and keeping us safe from its hazardous components are amazing! We need to give them less garbage by removing all the organic and recyclable materials from those trash bags. Don't just toss things when you aren't at home. Be sure to look for recycling and compost bins in your community. If there aren't any, maybe you can change that?!?

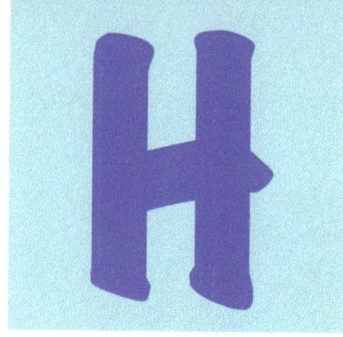

Home Composting

Our garbage is smaller when we take out the food waste and compost it. We challenge you to only throw away one bag of garbage in a week!

Do you see dumpsters or garbage cans overflowing in your neighborhood? What can you reuse, recycle, or compost to decrease your garbage each week?

The garbage that gets picked up from a home is called 'municipal solid waste' or MSW. Most of what the normal American family puts in the garbage bin doesn't need to go to a landfill. Let's see how low you can get your MSW by starting to compost.

Start composting by tracking how much organic waste is thrown away in 1 week. This is called a waste audit. First, place a container with a lid next to the garbage can. Instead of throwing away anything that can be composted (food, napkins, tissues, wood chips, paper towels, lint from the dryer, hair from your shower drain, etc.) collect it in the container. At the end of the week you will be able to see the volume of organic materials that would have been thrown away.

A lot of the items that end up in the MSW are recyclable or compostable.

When finding a bin to compost in, make sure it can hold more than the amount of materials collected in one week. Read on to letter O to for more information about types of compost bins.

Did you notice that the garbage isn't as stinky when the compostable products were put into the container? Isn't it amazing what happens when you take the compostable material out of the trash!

At the end of the week, see how well you did at reducing your trash! Put on some gloves and open up the garbage bags over an easily cleaned area, like an old shower curtain. Sort the items based on what they are made of.

Do you have a pile of plastic that could be recycled? Are there items that you could find a reusable product instead of a disposable product to purchase? There are places to recycle disposable razors, styrofoam, and contact lenses. With a little research, you can decrease your MSW by finding ways to compost and recycle that work for your family, school, or job site.

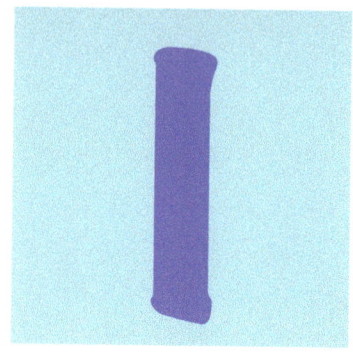

Invertebrate

There are so many unique invertebrates, no backbone, that live in the compost. Seeing evidence of these bugs in compost shows that it is a healthy ecosystem.

Carabid Beetle

Beetle Larva

Centipede

Rove Beetle

Fly

Worm

Mites

When stirring your compost watch for different invertebrates as the waste becomes awesome dirt.

Arthropods and annelids are invertebrates that eat organic matter after the thermophilic microorganisms have cooled down in the compost pile. Various types of springtails, mites, ants, beetles, centipedes, millipedes, snails, slugs, worms, and flies feast on each other and your waste in the ecosystem created in the compost bin.

In the Kingdom Animalia figure above, the blue branches are all invertebrates. These include the Phylum Annelida or annelids, which are types of segmented worms. Earthworms are a type of annelid that can be found in soil almost anywhere, especially the compost.

The Phylum Arthropoda includes the animals we call insects, spiders, centipedes, and millipedes. Did you see any in the compost?

Crustaceans are arthropods and can be thought of as 'water insects,' because they serve similar roles in the ocean as our insects and spiders do on land.

Many invertebrates can be found in your compost pile. Can you think of any vertebrates that might like to hang around it, too?

Journey

There is more than just the life cycle of a frog, butterfly, or pumpkin. There is the life cycle of compost with 5 basic steps:

Use compost to grow more food!

Harvest food

Eat food

Collect food waste

Compost food waste

The journey of organic material in a compost pile starts with thermophilic bacteria breaking down matter while heating the pile up to 160°F or higher. Next, mesophilic bacteria keep up the process and maintain the pile above ambient temperature. Finally, the pile will cool down and the invertebrates move in. A properly maintained compost pile can turn organic matter into compost or "black gold" in as little as four weeks. The finished compost can then be spread around plants to encourage growth. The plants will absorb and reuse the building blocks in the compost to make more plant cells. Animals, like humans, eat different parts of plants, like apples. The apple core can then be returned to compost to feed another round of plant growth.

The life cycle within the compost

Black gold used to make more food!

Mix collected food waste with a browns source to keep the smell down.

COMPOST

Thermophilic bacteria at work! Watch temperatures increases to 160°F.

The Greek 'thermos' means coming from heat! Thermophilic microorganisms HAVE to have it hot. They can also be found in hot springs or geothermal zones like Yellowstone National Park.

Mesophilic bacteria grow above ambient temperature chewing on the waste.

Meso comes from the word for middle. These bacteria thrive on the not-so-hot environment after the thermophilic action starts cooling down, but before all the heat is gone.

Finishing the compost.

Move in Day for Invertebrates!

The final stage is letting it sit. After another week or two, the invertebrates move on to greener pastures, the pH of the soil balances, and the black gold will be ready for you to put to work!

Once the compost pile has cooled to ambient temperature, the invertebrates move in and stay until the process is complete.

Keep track

It is fun to see the waste you are keeping out of the landfill by writing it down. It is also required by some state's laws.

Date	#s composted	#s recycled	#s in landfill	
12/13			83.5 lbs	
12/14	6.5 lbs	1 lb	66 lbs	Empty trash can 125.5 lbs
12/17	3.5 lbs	1 lb	66 lbs	Empty compost bucket = 1.5 lbs
12/18	4 lbs	1 lb	76.5 lbs	Empty recycle bin (blue) 8 lbs
12/19	5 lbs *2.5 lb milk	2 lbs	53.5 lbs	
1/7	4.5 lb *5 lb milk	2 lbs	45 lbs	
1/8	2.5 lbs *7 lb milk	2 lbs	52.5 lbs	

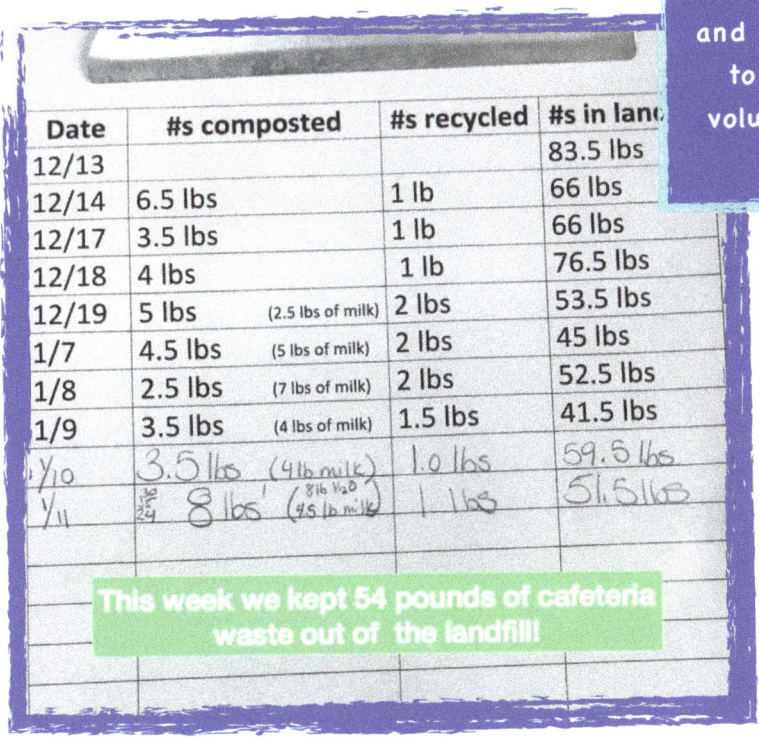

Date	#s composted		#s recycled	#s in land...
12/13				83.5 lbs
12/14	6.5 lbs		1 lb	66 lbs
12/17	3.5 lbs		1 lb	66 lbs
12/18	4 lbs		1 lb	76.5 lbs
12/19	5 lbs	(2.5 lbs of milk)	2 lbs	53.5 lbs
1/7	4.5 lbs	(5 lbs of milk)	2 lbs	45 lbs
1/8	2.5 lbs	(7 lbs of milk)	2 lbs	52.5 lbs
1/9	3.5 lbs	(4 lbs of milk)	1.5 lbs	41.5 lbs
1/10	3.5 lbs	(4 lb milk)	1.0 lbs	59.5 lbs
1/11	8 lbs	(45 lb milk)	1 lbs	51.5 lbs

This week we kept 54 pounds of cafeteria waste out of the landfill!

Keep track! Many states, like Colorado, have a limit to the amount of compost kept on site. Some cities also have rules about how much compost a location can have or what type of containment system you need to put it in. Having trouble finding your state's rules? Contact the county department of public health and environment.

If composting at school, be sure to write down how much food waste is collected each day. Not only does this keep the records needed for state requirements, but it is fun to add up how much waste is diverted from the landfill each week, month, and year! Estimate the compost by the volume of the buckets used (for example 5 gallons) or weigh it with a scale.

There are other ways to keep the cafeteria or home waste out of the landfill.

Does your school have a Share Table? This is a place you can put unopened items for others to eat if you are done with your meal.

Do you use disposable water bottles? Look at the labels. Many bottles have water from different municipalities or towns. Is their town's water any better than your water fountain? Could you use water bottles or cups to fill a glass instead of having disposable bottles?

Does any leftover water or milk end up in the garbage? By emptying these liquids into a watering can, you can water plants instead of adding to the garbage! Plants love the calcium from milk, especially strawberry plants.

Leachate

In a landfill, leachate liquid carries horrible, toxic pollution that must be trapped to prevent it from getting into our aquifers. In a compost pile, the leachate is so healthy it can be used to water plants.

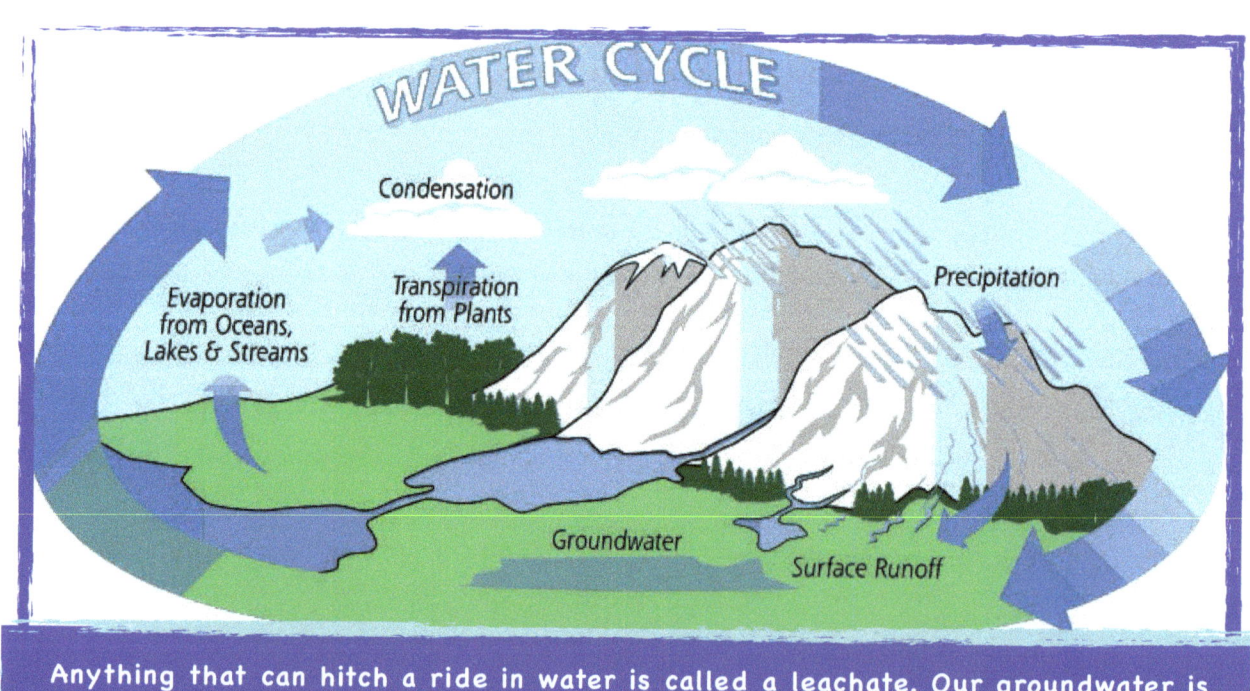

Anything that can hitch a ride in water is called a leachate. Our groundwater is full of minerals and other elements from the soil that swim along in our runoff.

Leachate is any liquid that passes through something picking up soluble pieces as it moves. Leachate travels through the compost pile and into the ground below it. Any contaminants in your compost pile could be incorporated into the leachate and end up in your soil or ground water. Keep paint, stains, medications, etc. out of the compost pile.

Coffee is a leachate from coffee beans!

Rain on landfills seeps through the garbage piles and becomes a toxic leachate. This leachate is hazardous for our groundwater. We are very lucky that landfills put a protective coat of clay and plastic under the huge piles of garbage to keep our ground water safe. We help by not putting toxic or hazardous chemicals into our garbages.

Residents in most cities can take hazardous household waste, like batteries, motor oil, old paint, CFL light bulbs that have mercury, fertilizers, pesticides, household cleaners, and many other hazardous items, to a drop off facility to be disposed of properly. Check at https://search.earth911.com/ to find a location near you. Ask a local pharmacist where or when a pill drop off is so unused medications can be taken care of correctly, too.

Don't toss those old electronics into the garbage either! Find an E-recycler or take them apart yourself to remake the device into something you can use. Old appliances, mattresses and furniture should be taken to a specialized facility to recycle. Colorado Springs has metal recyclers that will take old appliances to be recycled.

Be sure to look at what is being throwing away. If it is a chemical or metal, it could be hazardous. Be sure to take any of these items to the correct place for disposal. Help prevent fires in garbage trucks and prevent the landfill's leachate from being so toxic.

Batteries are the number one cause of fires in garbage trucks!

Microorganisms and Macroorganisms

There are five basic kingdoms: Monera, Protist, Fungi, Plant, and Animal Kingdoms. Monera, Protists, and some Fungus are single-celled microorganisms that are so tiny you need a microscope to see them. Macro-organisms are fungus, plants, and animals you can see with your eyes. Microorganisms and microorganisms break down our food waste in the compost.

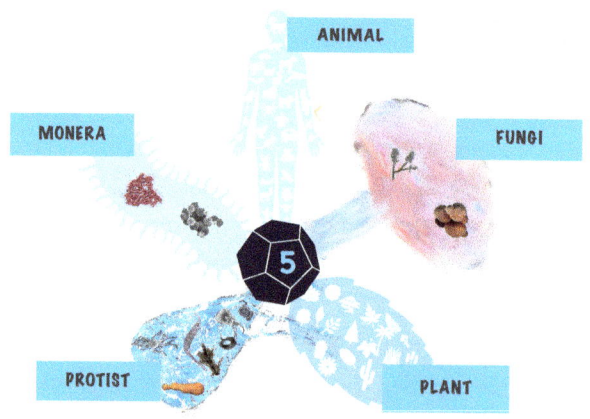

All living things are classified into one of the 5 kingdoms. All of the 5 kingdoms benefit from composting.

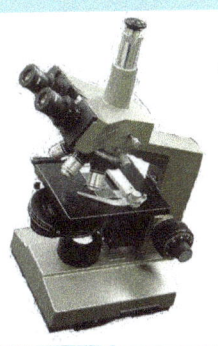

Microorganisms are entities that are not visible by the human eye. The kingdom Monera has many types of microscopic bacteria that are essential in the breakdown of material in the compost pile. The soil is also full of many microscopic Protists that can be found digesting our food waste in the compost piles.

Fungi can be both microscopic, like yeast or mold that is in the compost, or macroscopic, like a mushroom that is getting food from decomposing tree roots. Macroscopic organisms from the Animal kingdom that you can find in the compost include many invertebrates. Sometimes you might even see a scavenger animal, like a bird, mouse, or raccoon around the compost. Plants benefit from the compost being added back to the soil, where their roots can reabsorb all the yummy nutrients.
All 5 kingdoms are a part of composting!

This phase contrast microscope helps our human eyes see much smaller objects… the 'micro'scopic kind!

Light is moving in parallel lines until it comes to a lens, like the one shown here. The light that passes through a convex lens (L) converge at the Focal Point labeled F. After the light converges it moves away in non-parallel lines which leads to amplifying the picture to someone looking at it. This is how a magnifying glass works.

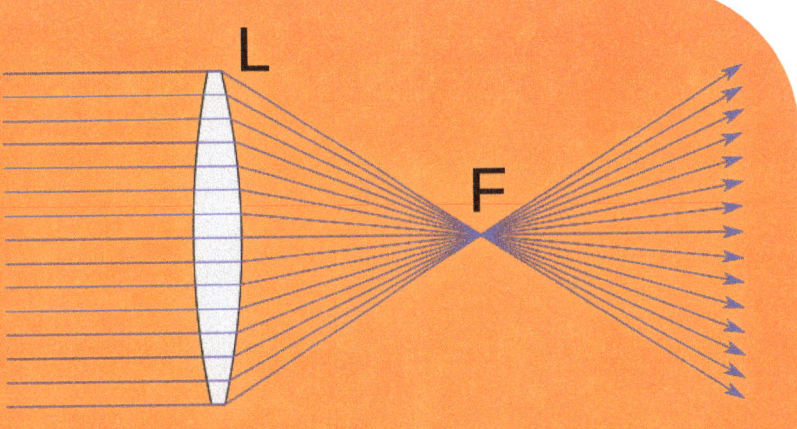

L

F

Microscopes use many different types of lenses to converge light in the best path so when you look through the eyepiece or ocular lens you see small things as large!

A magnifying glass is a great way to start noticing the microscopic details in a macroscopic object!

Nitrogen

Nitrogen is a main ingredient in the compost cycle. It is needed for plant and animal life to grow. That is why we need to recycle it instead of throwing it in the landfill.

Food waste is a great source of nitrogen. This is some food waste collected in 5-gallon buckets from the school cafeteria to go into the compost.

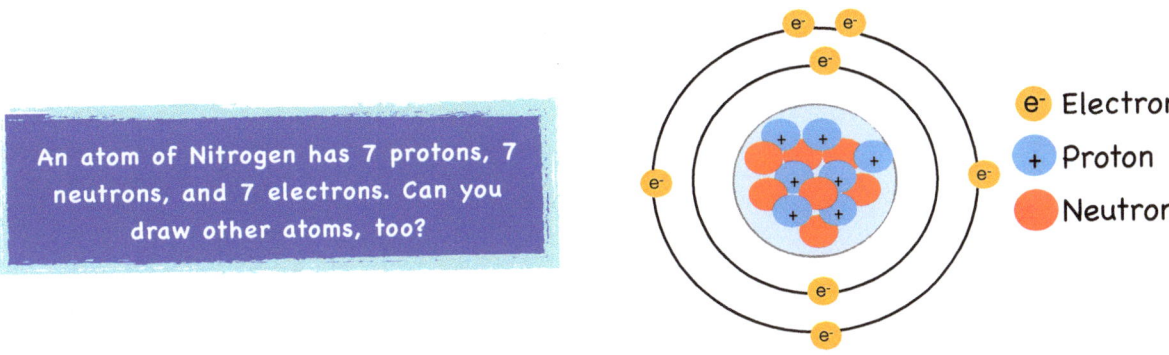

An atom of Nitrogen has 7 protons, 7 neutrons, and 7 electrons. Can you draw other atoms, too?

e⁻ Electron
+ Proton
Neutron

The nitrogen cycle on Earth keeps this element available for new protein synthesis in all organisms. Proteins are a type of macromolecule found in living things. Remember, biodegradable items like macromolecules can be broken down and reused by living things.

Nitrogen is found in fertilizers because is essential for plants to grow. Instead of using fertilizer, compost can be spread to return nitrogen to the soil. Organic materials with a lot of nitrogen are referred to as "greens," where as materials with more carbon are called "browns." Compost contains nitrogen and carbon that can be used again by growing plants.

Nitrogen is represented by an 'N' on the Periodic Table. The Periodic Table lists all the known individual elements on Earth, and their atomic properties. Human bodies are made primarily of hydrogen, oxygen, carbon, nitrogen, phosphorus, and sulfur. Can you find these elements?

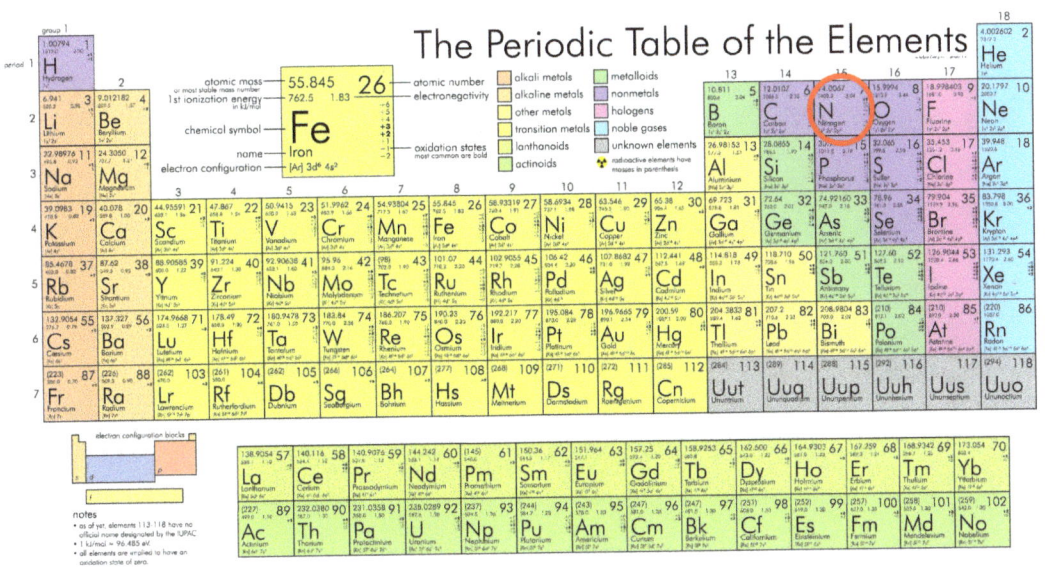

The Periodic Table of the Elements

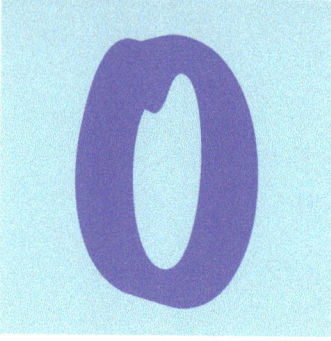

Organic Material

Organic material is something made from carbon that can completely decompose back to the Earth. Examples are: orange peels, banana peels, manure, hair, dryer lint, paper towels, chicken bones, and even peanut butter! Plants and compost love milk, too!

This plastic cup is not made of organic material. It will break up into smaller and smaller pieces, but the living organisms of the Earth won't be able to use it for food or energy.

Organic material for a compost pile is anything carbon-based. Organic matter comes from plant or animal parts and pieces. The Earth is dependent on cycles to reuse elements constantly, such as the carbon cycle, water cycle, and nitrogen cycle. We can use our compost bins to return our food and yard waste to the soil. Bins, tumblers, or stacks are common containers to store your organic material as it composts. But just leaving it in a pile will allow for decomposition, it just takes longer.

There are many different compost systems. Depending on how much space is available, consider building a bin system or purchasing a compost tumbler. Tumblers are nice if space is limited. Some even have a sifting mechanism inside to separate the composted material from the newly added waste. A 3-bin system is convenient if there is room: one bin is for new waste, one bin has the older waste that is finishing the composting process, and the third bin is empty to transfer compost into when mixing weekly. Trenching can be done if there is a large area. Dig a trench and put organic waste into it, cover, and let nature do its thing.

Mix your carbon (browns) and nitrogen (greens) rich materials into the compost system. Keep the compost moist and stir it weekly to have the fastest results. Sift the finished compost to return large pieces to the bin to continue breaking down and be able to put "black gold" in your soil!

There are many different types of composting systems. Find the type that holds what you have and best fits your lifestyle.

Plastic

Plastic use exploded in 1945 and has grown since. Plastic is now in drinking water, beaches, most rivers and lakes, and even found in our blood. All the small pieces aren't able to be recycled. Manufacturers and restaurants should sell their products in compostable containers. You could invent new Earth friendly containers and make millions!

Plastics do not break down into basic elements again. They become smaller pieces of plastic known as microplastic or nanoplastic. Plastics are made from petroleum and are long chains of carbon. Microbes are unable to utilize them as a food source because of the chemical process done to the carbon chains by humans is too strong for bacteria to mess with.

Small plastic beads are even used as part of facial exfoliants. These little balls of plastic wash into the city's waste water and can't be removed. Since water runs down-hill, any plastics that are in your waste water will end up flowing down stream to neighboring towns. Reducing our plastic use will prevent microplastics from contaminating our soil and water.

One of the most common plastics is called polyethylene terephthalate (PET). PET is used in polyester shirts and single-use water bottles. It has number 1 as its resin identification code, which is the number you will find in the recycle arrow triangle on plastic products. There are 7 types of plastic and not all can be recycled. Be sure to check with your recycling center to make sure only the plastics they accept are in your bin!

OOPS!

Get something in the compost that doesn't belong? It's OK! Just remove the non-biodegradable item and let the rest turn into awesome dirt.

Do you buy a product in a container that can't be recycled or reused? Contact the company and see if they will change the packaging. Or buy a product in a compostable, reusable, or recyclable container.

Quatro

Quatro means four in Portuguese. There are four basic components in compost: nitrogen, carbon, water, and oxygen. When mixed together in the right ratio they create the perfect compost pile.

greens (nitrogen) + browns (carbon) + water + oxygen = awesome dirt (aka compost)

In Spanish it is spelled cuatro.

Your compost bins are a quad-RAD-ic equation of decomposition! Quad means four in Latin, just like quatro means four in Portuguese. The four components to have quick, nice smelling compost are: browns, greens, water and oxygen. The fastest way to make awesome dirt is to mix the right ratio of carbon (browns) and nitrogen (greens), keep the pile moist with water, and mix regularly to incorporate oxygen.

Finding the mix of browns and greens varies for every compost pile. Because you are using organic material to make compost, each of the items will have some amount of carbon and nitrogen. Don't be afraid! You don't have to be a mathematician to figure this out! You can do really complex math if you like, but it is not required. The minimal thing to remember is to have more carbon rich, brown, material than nitrogen rich, green, material and your compost won't be rank. What happens if you put in too much brown material? The composting process will be slower. Not a bad consequence for not wanting to do math, right?

	Material Type	C:N ratio
Browns	straw	75:1
	dry leaves	60:1
	newspaper	175:1
	sawdust	325:1
	wood chips	400:1
Greens	garden waste	20:1
	manure	15:1
	food scraps	20:1
	vegetable remnants	25:1

Items like food waste have less carbon available than dried things, like paper and sawdust. The table shows common C:N ratios of items you may have to compost. As you can see, food scraps still have roughly twenty carbon molecules to one nitrogen molecule (20:1 ratio). The general guideline is a final 30:1 ratio of carbon rich materials (browns) to nitrogen rich materials (greens). Depending on the type of browns, there will be a larger volume than others. For example, you will use more shredded paper, by volume, then you would use sawdust or wood shavings.

Shredded materials will compost faster. The more surface area available for the microbes and invertebrates to snack on, the faster they will chew it up. So shred that paper, chip that wood, smash those old pumpkins, and crush those egg shells before tossing in the compost bin.

Reduce

Composting is a great way to turn food waste into dirt, but having less waste to compost is better. How can you decrease the amount of organic material to compost in your life?

Many paper products can be replaced with washable cloth alternatives!

Eat what's in your fridge! Don't let those forgotten leftovers go to waste.

Did you know that composting food waste is almost the last resort before going to the landfill? The Wasted Food Scale from the Environmental Protection Agency is a diagram of how to try to stop food from ending up in the landfills.

https://www.epa.gov/sustainable-management-food/wasted-food-scale

First is Preventing Food Waste. Cafeteria staff work hard to make sure they have enough food for you, and so do your parents. Help them know how much to make by not taking food you won't eat. Be an active participant in making your lunches or other meals so you eat what is prepared.

Next we can Donate or Upcycle. Ask if you can put a Share Table in your cafeteria. If someone has unopened food (like you would get from the grocery store, not a sandwich with a bite out of it!) you can put it on the Share Table for someone that may still be hungry. Also, you can see if the leftover food from the cafeteria can be donated to a local food bank or shelter.

Animals like pigs, chickens, and goats love our leftovers! Check on local regulations if you can feed food scraps to animals. Some farmers can leave food unharvested, to let it decompose back into the field instead of stores putting it in the landfill later.

Composting is a great way to make sure your food waste isn't ending up in a landfill! Anyone can do it, anywhere you are!

Share tables don't need to be fancy. It's just a spot to put something you aren't going to eat that someone else might like.

Some industries use different types of food waste to make energy. Look in your community and see if there are any biofuel plants (anaerobic digestion plants) that could use some extra food waste.

Which food waste reduction option is best for you?

Safety

It is important to be safe around the compost pile. Wash your hands immediately after taking care of the pile. Wear coveralls or change clothes after churning or mixing the pile.

Scrub your hands for at least 30 seconds, which is the time it takes to sing "Happy Birthday" to yourself twice. Try it the next time you wash your hands!

The compost pile is full of various microbes. When mixing or stirring the compost, wear gloves. A mask or face covering can help to prevent inhalation of spores that become airborne. When done with the compost, wash your hands! After stirring the compost pile, avoid wearing those clothes and shoes around your house or school.

Check the compost weekly. If the compost pile heats up and dries out, it can start a fire! Some tumblers are more prone to drying and starting fires, but fires can't start if the pile is kept moist. The compost needs to heat up to keep it safe. Compost that does not heat up has the potential to spread disease causing microorganisms. It is important to remember that compost is an active ecosystem full of microbes and although we don't need to be scared of it, take precautions to ensure that anyone around the compost stays safe. This includes making sure the compost heats up and follow safety recommendations when working with compost.

When stirring the compost wear clothes you can remove and wash. I use an old cotton bandana as a mask and ALWAYS put on my gardening gloves that have been designated for composting. Make sure any cuts or scrapes are covered with a bandage or waterproof tape.

Keep your compost pile moist and check on it at least weekly to prevent fires.

Thermophilic

Heat is a needed factor in the compost pile. Thermophilic bacteria heat up the compost pile as they break down, or decompose, food. This also kills germs that could harm us.

Compost is hot! The first weeks of decomposition are exothermic. Temperatures should reach 135°F – 160°F.

Not only do thermophilic bacteria create heat, but they are able to live in hot places, like this hot spring in Yellowstone National Park. Finding a steaming location is a great place to learn about thermophilic microorganisms.

Some bacteria are thermophilic bacteria, which means heat is created when actively reproducing. The thermophilic bacteria in the compost pile are the first to start decomposing and will raise the temperature to over 130°F. Not only is this great for those thermophilic bacteria to grow, but when the compost pile reaches 135°F – 160°F for 2-3 days pathogenic organisms that may be present are killed. If the temperature goes above 160°F your little thermophilic buddies may not survive. Unattended compost piles with high nitrogen content can easily get well above 160°F, dry out, and become a fire risk.

Stirring the compost pile prevents it from getting a very hot core and warm sides which could lead to fire troubles and slows down the overall composting process.

In the first weeks of composting, you can see steam when you stir it!

Universal

The most amazing thing about our planet is that all living organisms are made of universal material that can be recycled and decomposed back to nature to be used again.

Would you use flour to make a cake or paper mache?

Flour can be mixed with ingredients to make a cake OR it can be used to make paper mache for a piñata. Like using flour to make different things, the elements of the periodic table can combine to form the many parts and pieces that make life on earth possible.

The elemental components that make up proteins, DNA, membranes, and cells are universal. These molecules can be broken down and repurposed over and over again. This process is vital to the recycling of available nutrients on Earth. Humans can use chemistry to manipulate these elements into configurations that life on Earth cannot reuse, like plastics. Living organisms keep reusing the building blocks of life over and over again, unlike products, like plastics, that are difficult to reuse or not-reusable by living organisms at all.

Vermiculture

There are three types of worms:
flatworms -like flukes-,
roundworms -like nematodes-,
and annelids -like earthworms-.
Annelids like to make a home in the compost pile.

Red Wigglers are annelids sold in gardening stores.

Earthworms and red wigglers are common annelids found in compost.

Vermiculture is the act of using an enclosed system with worms to decompose matter. Vermiculture towers can be purchased or made and kept in your school, home, or office with very little upkeep. The worms eat your food waste and paper while leaving casings that are very nutrient rich for the garden and plants. When it is time to use the dirt created in the vermiculture system, there are two easy methods to collect the majority of the worms again:

1. Pile the output in a well lit location. The worms will gravitate into the pile and away from the light. Remove the top few inches of dirt or deep enough that you don't grab any worms. Leave the pile in the light and the worms will go deeper into the pile again. Repeat this until all the worms are in a small amount of dirt that can be returned to the vermiculture system.

2. Move the black, casing filed dirt to one side of the bin. Fill the now open side with moist shredded paper or cardboard and some food scraps if you have some available. The worms will gravitate to the new food source and you can start pulling the finished material out in a few days. Place in a well lit location to sift gently through and make sure eggs or other worm stages are recovered.

Take your awesome dirt to the garden, grass, or a plant that you want to give some extra love to, or use it to make compost tea!

In an outdoor composting system, worms will work their way into the compost after the temperatures have cooled down. Sifting the compost after it is finished can help recover these new invertebrate friends and return them to the compost system to start the process of breaking down organic matter again.

Web of Life

All species are connected in the Web of Life. In order to survive, we need to care for the tiniest and largest species. If one organism dies, the web falls apart and the other organisms are effected, too!

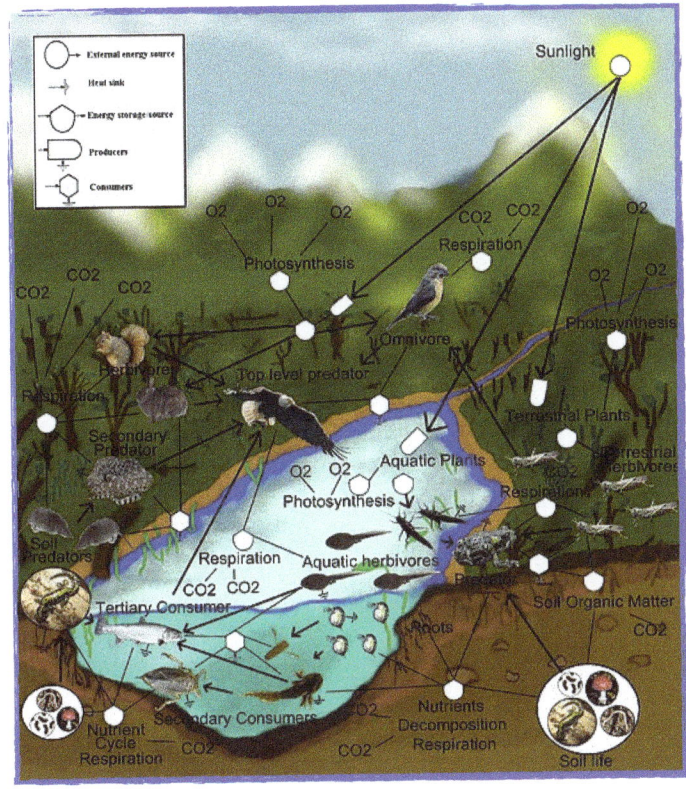

Sunlight

Photosynthesis

Respiration

O_2 O_2 O_2

CO_2 CO_2

O_2 O_2

O_2

CO_2 CO_2 CO_2

Photosynthesis

Omnivore

Herbivore

Top level predator

Respiration

Secondary Predator

Terrestrial Plants

Aquatic Plants

O_2 O_2

Terrestrial herbivores

Photosynthesis

Respiration

Soil Predators

Respiration

CO_2 CO_2

Aquatic herbivores

Predator

Soil Organic Matter

Tertiary Consumer

Roots

CO_2

Nutrient Cycle Respiration

CO_2

Secondary Consumers

CO_2

CO_2

Nutrients Decomposition Respiration

Soil life

Ecologists are biologists that study how living things interact with each other and the environment.

There are four laws of ecology:

1. Everything is connected to everything else. There is one ecosphere for all living organisms and what affects one, affects all.

2. Everything must go somewhere. There is no 'waste' in nature and there is no 'away' to which things can be thrown.

3. Nature knows best.

4. There is no such thing as a free lunch. Exploitation of nature will inevitably involve the conversion of resources from useful to useless forms.

The compost has its own ecosphere or web of life.

Primary consumers (bacteria, fungus, actinomycetes, nematodes, some mites, snails, slugs, earthworms, millipedes, sowbugs and white worms) feast on the organic residues.

Secondary consumers (springtails, some mites, feather-winged beetles, nematodes, protozoa, rotifera, and soil flatworms) eat primary consumers.

Tertiary consumers (centipedes, predatory mites, rove beetles, fomicid ants, and carabid beetles) go after secondary consumers of the compost pile. These amazing invertebrates then become energy sources for spiders, birds, reptiles, amphibians, and the chain of energy continues into larger and larger animals.

Any living thing that dies or any product from those living things starts the cycle all over again with composting or decomposition. There is no waste in nature. The plants and animals do not have an 'away' to throw things or garbage trucks to pick it up. The web of life keeps the elements and molecules that make up living things recycling through our Earth's systems to continuously create the next generation from the last.

Xerophobous

A compost pile won't survive without water. You need to keep it moist so you can observe fungus or molds!

The compost is a living ecosystem so water is a must! Dehydration is what keeps food from spoiling and that is the opposite of what is going on while composting.

In order to optimize the time it takes for your compost to turn from waste to new dirt, keep it moist. Many of the organisms in the compost pile are xerophobous, unable to survive drought or drying conditions. By keeping the compost pile aerated and moist, the microbes stay alive and the time it takes for material to decompose decreases.

Keeping your compost moist not only helps the process finish quickly, but it helps prevent fires. Remember that compost can reach temperatures over 160°F and without water it could be asking for trouble. Keeping compost bins leaned up against structures, like houses, is not recommended because of the possible fire risk.

Dehydration is a common food preservation method because microbes can't grow without water. Don't preserve your compost pile! Keep it moist.

Which will mold? The wet or the dry chiles? Mold is a good thing when composting!

YOU can do this!

Whatever your biome is, it needs you to reduce waste! We challenge you to reduce your school trash to the size of one 5 gallon bucket per day and reduce your home trash to one bag a week! Challenge your family, your neighborhood, your school, and your city. Can you challenge another school, maybe even in another country? Write a letter to them. Tell us about your progress at www.thinkingdirty.org . Then challenge another school in the world! Let's do this!

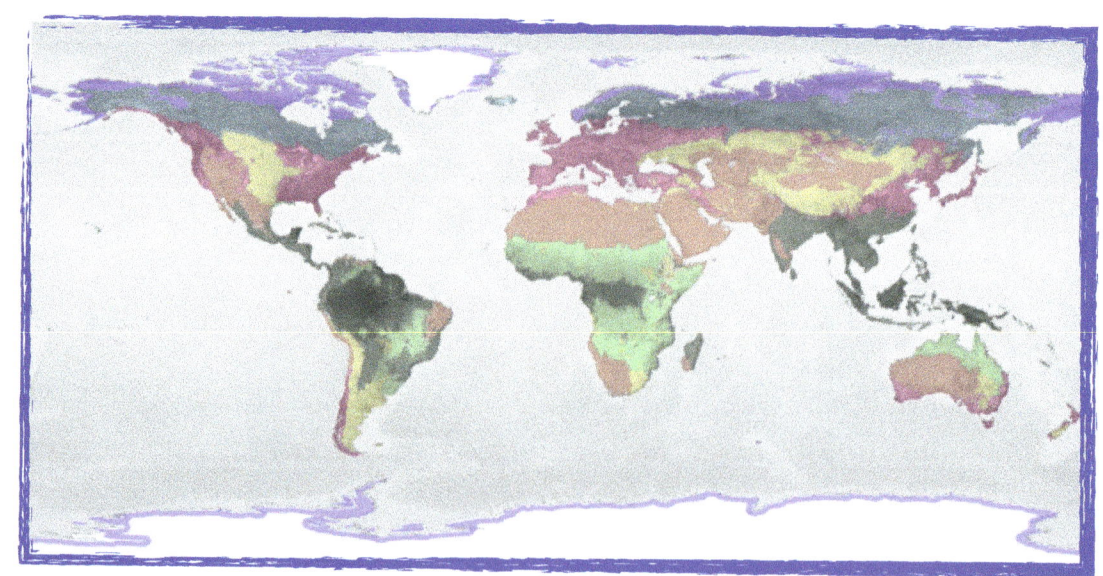

Find your biome on this interactive map at: https://earthobservatory.nasa.gov/biome

It's go time! Green light your composting efforts and start your waste reduction journey.

You can compost. The only way to start composting is to just do it! No one gets it perfect the first time. You continue to experiment with the organic waste to find the best mix of browns and greens, and how often to mix and water it. Every compost pile is different, but even leaving it in a heap will allow for decomposition. Before you know it you will have black gold. The biome you live in will love getting these nutrients to keep the cycles of life on Earth rolling.

Think your compost is not savable?! Trench it! Dig a trench 1-2 feet deep in a flower bed or garden, pour the compost in, and cover with dirt. Earth's amazing soil is already full of the microbes and invertebrates to turn the trench into compost. Then take what you learned and start another compost bin utilizing your new knowledge. Albert Einstein once said, "Anyone who has never made a mistake has never tried anything new." Failing in science is inevitable, but we dust off our hands, learn from our mistakes, and keep going.

Join the Compost Challenge! Get information and send your before and after pictures!

Zigzag

Starting to compost will have ups and downs like the zigzag of a rollercoaster. Keep your eye on the prize and you will have a system worked out to make awesome dirt!

Pursue some path, however narrow and crooked, in which you can walk with love and reverence. ~Henry David Thoreau~

If starting to compost was easy, everyone would do it already. Once the process is started, the journey from organic waste to awesome dirt is not a straight line. You can plan to have a day when the compost is smelly and more browns need to be mixed in. You may find out that a neighborhood raccoon has made its way into the bins. Perhaps there are strict regulations in place in your area and you will need to get approval from a governing body to start composting. These can be frustrating at times, but always remember that in the end the organic waste will decompose leaving wonderful, nutritious dirt for your plants. There is no perfect compost system and each system is different because we all have different amounts or types of waste. Take the zigzag, rollercoaster ride that will be your composting project on with a positive outlook and patience.

They say it takes a community to raise a child. It can also take a community to raise some awesome dirt. There are many online resources with people who have been in your shoes and would love to help trouble shoot any issues you may have. We are available through www.thinkingdirty.org or contact your local extension agent or sustainability program. Look at other schools or businesses in your town or state and ask for help. The composting community is out there and would like your project to be a success.

Acknowledgements

The authors are indebted to many places for their time, energy, and mad skills.

Mary would like to thank:

Thank you to WM (formerly Waste Management) in Colorado Springs for all the wonderful information about their composting site, the tours, pictures, and having the only class III composting facility currently in Colorado Springs!

Thank you to Buena Vista Elementary Montessori School for being brave enough to take on composting cafeteria food. So much gratitude to all the teachers, especially Carrie Delius for everything she did to spearhead the project. And thank you to Ms. Tami for going above and beyond to remove the styrofoam trays and plastic sporks from the lunch program!

Thank you to Zach Brown at Brown's Greens for the many conversations on composting in Colorado Springs.

Jill would like to thank:

Thank you to Dr. Maria Montessori, the first female doctor in Italy, who advocated for children's education and inspired us to bring peace and tranquility to our Earth. Dr. Montessori has a quote "Bring the world to the child, or the child to the world." Our mission is to rebalance our world so that this is possible.

Thank you to Rose Seeger, who owns Green City Resources and installs green gardens on the tops of Cincinnati's hospitals and other beautiful buildings. She brings joy to so many.

I would like to thank Kim Hay, Carole Ann O'Neil, Terry Hartnett, George Schmidt, Mimi Jones, Robin Prosise, Shari Kasten, and Karen Kight. These wonderful people stood by me and inspired me to be the teacher that I am today!

* 9 7